TEDDY BEARS

AN ANTHOLOGY OF VERSE & PROSE

TEDDY BEARS

AN ANTHOLOGY OF VERSE & PROSE

This edition published in 1998 by Hermes House
27 West 20th Street, New York, NY 10011

HERMES HOUSE books are available for bulk purchase for sales promotion
and for premium use. For details, write or call the sales director,
Hermes House, 27 West 20th Street, New York, NY 10011;
(800) 354-9657

Hermes House is an imprint of
Anness Publishing Limited

ISBN 1 84038 075 6

Publisher: Joanna Lorenz
Project Editor: Joanne Rippin
Consultant Editor: Anne Gatti
Designer: Kit Johnson

Printed in Singapore by Star Standard Industries Pte. Ltd.

1 3 5 7 9 10 8 6 4 2

Contents

Chapter 1

The History of the Teddy Bear

During the first half of the Twentieth Century it was found that the subtle appeal of the Teddy Bear was so endearing and enduring that the Teddy has become the lasting symbol of childhood, and consequently outlived all other mascot animals. It is now realised that anybody who thinks the Teddy Bear is just a cuddly toy and nothing more is very much mistaken. There is far more to Teddy than meets the eye. For there is now ample evidence to show that the Teddy Bear gives solace and enjoyment to people of all ages and both sexes. So much so, in fact, that this takes it right out of the classification of a soft toy.

Colonel Robert Henderson,
the great Scottish collector of teddy bears

On 14 November, 1902 the President of the USA, Mr Theodore "Teddy" Roosevelt, was attending a hunt for grizzly bears in the state of Mississippi. After several days of unsuccessful hunting his embarrassed hosts produced a young captive bear for the president to shoot at. Roosevelt, who was well known as a sporting gentleman, contemptuously refused to kill such easy prey. Clifford Berryman, a cartoonist on the *Washington Star*, portrayed the incident in a cartoon in which the grizzly bear cub on a rope bore more resemblance to a toy than a real bear.

The cartoon in the *Washington Star* was seen by a toy maker from Russia, living in America, Mr Morris Michtom. He and his wife were charmed by this story of their humane president and devised a cuddly toy which looked something like the animal Berryman drew. They called it "Teddy's Bear", and its success was immediate. Soon the Michtoms were selling as many teddies as they could make.

SUGAR

In Germany, at around the same time, at the beginning of the 1900s, a woman called Margarete Steiff was making her living as a seamstress, determined to be financially independent despite her lasting disability from childhood polio. She had a great love for children and would make them small animals out of cloth left over from dressmaking.

Margarete's nephew, Richard Steiff, had been studying bears at Stuttgart Zoo and at his insistence Margaret designed a toy bear which they took to the Leipzig Toy Fair. He was called "Friend Petz" and his success at the fair was immediate with an order from America for 3,000 copies.

Whether the teddy was invented in America or Germany it was an extraordinary success. In Britain the teddy was affectionately linked with Edward, the Prince of Wales, who not only shared the same name with the little bears, but also had a similarly portly figure. With royal and presidential seals of approval teddies were on their way to becoming the world's most famous and beloved toy.

The Teddy Bear has come to stay, so perfectly is his grizzly exterior adapted to fitting into the many chubby arms which are extended to him. He is not only bear-like enough to lift him above juvenile criticism but he is possessed of those semi-human attributes which fit him eminently for youthful companionship. He is every inch a bear and yet he certainly embodies exactly the doll qualities which are demanded by the child of today. He is well-made and set up. His head really turns round and his legs are nicely adjustable. He has moreover that precious gift of true adaptability; he can be made to crawl, climb, stand or sit and in each pose he is not only delightfully himself but he also suggests to the imaginative owner whatever special being his fancy would have his teddy personify.

Caroline Tickner
New England Magazine, 1907

Chapter 2

Loving a Bear

I'm really Edward George St Clare
Aubrey Adolphus de la Bear
Son and heir of the Baron Bear
But you may call me Teddy Bear.
So please, squeeze me, I don't care.

Inscribed underneath a drawing
of a brown velvet teddy bear in an
autograph album belonging to Mrs James,
Chartham, near Canterbury, England,
6 February 1912

Egbert & Me

At night when I lie fast asleep
My teddy, Egbert, wakes,
And sits upon my counterpane
Until the morning breaks:
He likes to see I get my rest,
For everybody's sakes,
So if the pirates smash the door
To steal away my toys
He fights them off with dirk and sword
But very little noise.

And if the one-eyed bogey-man
Comes breaking down the wall
He scares him off by looking fierce
But makes no sound at all.
And if the wailing ghost flies down
The chimney like a bird
He blows him back with mighty breaths
That simply can't be heard.

When, after slumbering peacefully,
I open up my eyes,
I see the sun come shining in
And find to my surprise
That Egbert's lost another ear
Upon some enterprise.
But when I ask him what he's done
He just looks smug and wise . . .

Anon

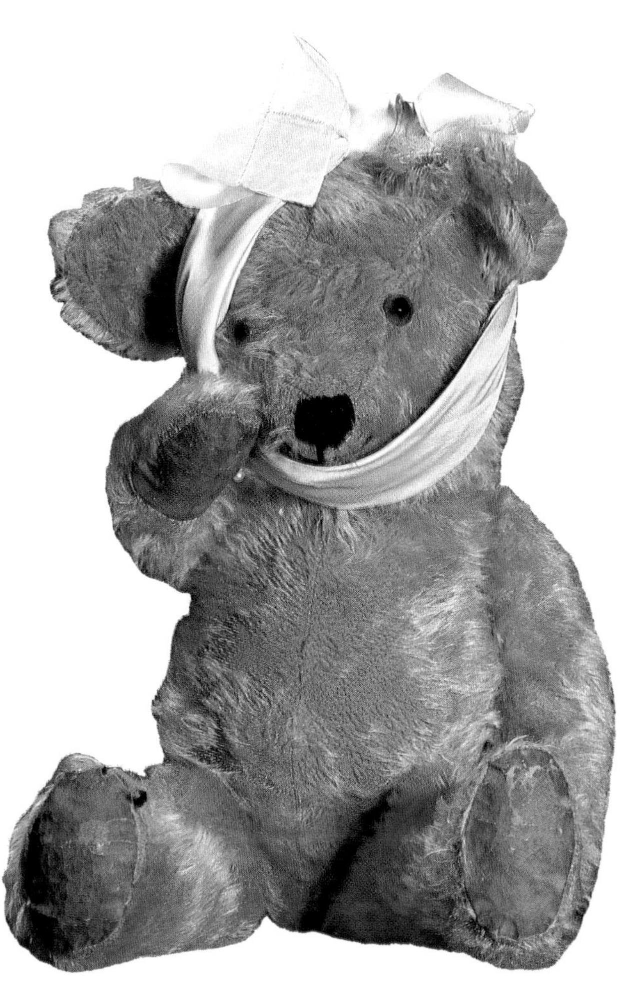

TEDDY WORSHIP

All twerps who worship teddy bears,
Pathetic wimps and drips,
Who cast them soppy moist-eyed stares
And kiss them on the lips,
Those nerds who huggle Fatty Bum
And snuggle up with Gus
And fondle Lamb and Bubble Gum
Bring shame to all of us.

What is this cult of teddy love,
This arctophiliac fit
That drags the high from up above
Into the sissy pit?
What is this craze that overtakes
The sanest of the sane
And leaves them coo-ing dribble lakes
In baby dreams again?

It's sad, I said to Teddy Tub,
My friend, the other day
That grown-ups should simp and blub
And carry on this way:
You wouldn't catch me being so crass
I said, you know it's true –
It's just a special friendship as
I cuddle up to *you*.

Pauline Jones

Go to sleep, my Teddy Bear,
Close your little button eyes,
And let me smooth your hair.
It feels so soft and silky that,
I'd love to cuddle down by you,
So,
Go to sleep, my darling Teddy Bear.

Anon, lullaby

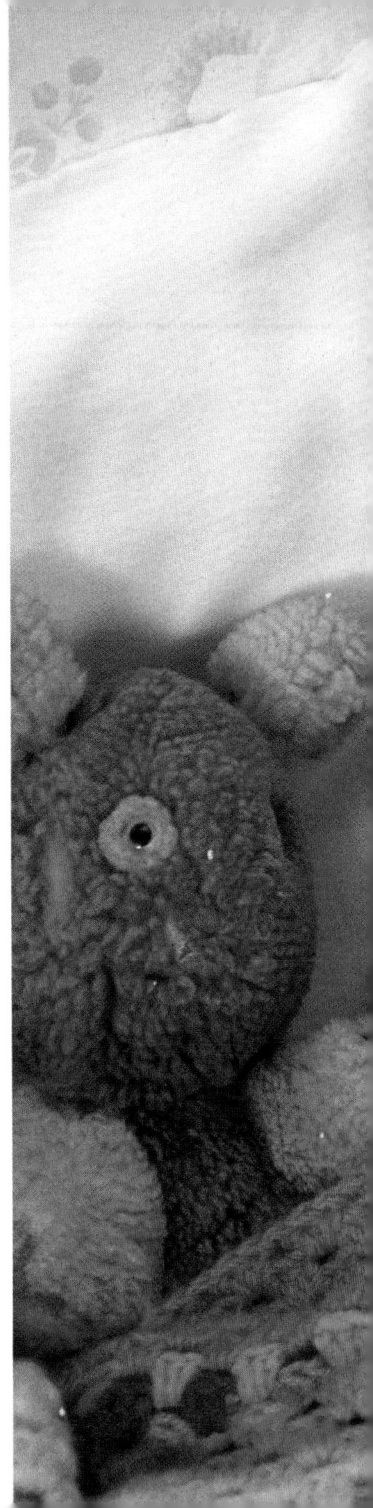

First Love

My teddy, Grumpkin, calls the tune,
For her I wrestle down the moon,
Fetch winter grapes, a dress of lace,
And eat up every wrinkled prune.

She sets the rules and I concur,
I count the ocean's drops for her,
Fight fearsome trolls, do forward rolls,
And, if she lets me, comb her fur.

When I detect her vaguest mood
I sail the seas in vessels crude
To fetch her treats, exotic sweets,
And potions rich with dreams imbued.

I strain to serve her slightest whim
And travel to hell's hottest rim
To brave the fire without a lyre
And pluck the Devil's nose from him.

And if you ask me, Why do so?
It's difficult for me to know.

Paula Ness

As teddy bears come close to their 100th birthday the demand for restorers and teddy hospitals grows larger. In 1972 two veterinary surgeons wrote a survey entitled Some Observations of the Diseases of Brunus Edwardii.

For more than a century, this species has been commonly kept in homes in the United Kingdom and other countries in Europe and North America. Although there have been numerous publications concerning the behaviour of individuals (Milne, 1924; 1928; *Daily Express* numerous editions), there have been no serious scientific contributions and a careful search of the literature, using abstracting journals and computerized data retrieval systems, has failed to reveal any comprehensive survey of the diseases of these creatures. A few of these previous publications include references to certain disease syndromes and Milne (1928) refers to obesity associated with the excessive intake of honey, and to psychological disturbances associated with territorial disputes with Tiggers, Heffalumps and even small children. One publication (Bond, 1958), concerning a certain individual known as Paddington, refers to the animal receiving treatment from medical practitioners without a veterinary qualification. These records emphasize two disturbing factors, firstly, the obvious need for treatment of diseased individuals and secondly, the infringement of the Veterinary Surgeons Act of 1966 that would presumably be involved if such animals were treated by any person not on the Veterinary Register . . .

Case 1: A six-month-old bear, owned by a four-year-old male, was found to be suffering from acute dyslalia, torticollis and loss of one lower limb. The general condition of the animal was good, with a normal thick pelage. The injury had been the result of disputed ownership. Treatment of the torticollis by manipulative therapy, and surgical replacement of the limb, were uncomplicated. The dumbness was the result of a ruptured acoustic membrane, and complete renewal of the voice box was necessary. This involved laparotomy, removal of the damaged organ from its surrounding viscera, and careful positioning of the replacement so that the acoustic membrane faced ventrally to prevent the development of muffled speech.

Case 2: A young bear owned by a child of six months was found to be suffering from "soggy ear" when removed from the owner's cot one morning. Oedema of the pinna was a commonly recurring condition in bears belonging to children under 18 months, who slept with an ear firmly clamped in their mouths. Treatment consisted in removal from the owner, lavage and drying in an airing cupboard.

DK Blackmore BSc PhD FRCVS, DG Owen MSc and CM Young, MA, VetMB, MRCVS from *The Veterinary Record*, Whittington Press, 1972

Chapter 3

Life with a Bear

Mr Archibald Ormsby-Gore has been with me as long as I can remember, he is about a foot high when he is sitting down and is very patched. His eyes are wool, his ears and nose are of some kind of cloth. Originally he was of golden fur but this only survives on his back and behind. He is very Protestant looking.

John Betjeman on his teddy

Safe were those evenings of the pre-war world
When firelight shone on green linoleum;
I heard the church bells hollowing out the sky,
Deep beyond deep, like never-ending stars,
And turned to Archibald, my safe old bear,
Whose woollen eyes looked sad or glad at me,
Whose ample forehead I could wet with tears,
Whose half-moon ears received my confidence,
Who made me laugh, who never let me down.
I used to wait for hours to see him move,
Convinced that he could breathe. One dreadful day
They hid him from me as a punishment:
Sometimes the desolation of that loss
Comes back to me and I must go upstairs
To see him in the sawdust, so to speak,
Safe and returned to his idolater.

John Betjeman from
Summoned by Bells (1960)

One of literature's most famous bears is Aloysius, Sebastian Flyte's companion in Evelyn Waugh's Brideshead Revisited. *This is an account of when Charles Ryder, the narrator, first glimpses Sebastian in Oxford in the 1920s.*

My first sight of him was in the door of Germer's and on that occasion I was struck less by his looks than by the fact that he was carrying a large teddy-bear.

"That," said the barber as I took his chair, "was Lord Sebastian Flyte. A *most* amusing young gentleman."

"Apparently," I said coldly.

"The Marquis of Marchmain's second boy . . . What do you suppose Lord Sebastian wanted? A hair-brush for his teddy-bear; it was to have very stiff bristles, *not* Lord Sebastian said, to brush him with, but to threaten him with a spanking when he was sulky. He bought a very nice one with an ivory back and he's having 'Aloysius' engraved on it – that's the bear's name."

Evelyn Waugh, *Brideshead Revisited*, 1945

In these modern days it has become unchic to relay or even retain any surviving relics of childhood character. I deplore this, as it seems to me, in this pressurized and atomic era, almost essential to have something to cling to which will remind one of the days when life appeared to be so complicated. Days which had been constructed entirely for one's enjoyment and entertainment. Meals were set in front of one as if by magic and one not only had no money problems but one never had to be in a Certain Place at a Certain Time. Or if one did Certain People took one there. Above all one was aware that there was always somebody to tell one's joys and sorrows to and, in far more cases than I had hitherto realized, this someone was Teddy Bear.

Peter Bull, *Bear With Me*, Hutchinson 1969

A Boy's Verse: To My Teddy

When I was only four days old
You came to live with me
And gave me all the love you hold
Quite unreservedly.

Oh Teddy Bear I still love thee
As much as I did then,
Though now we both are forty-
  three
And very nearly men.

Anon

My Teddy Bear

Written to Commemorate Teddy Bear's
75th Birthday

He sits upon his pillowed throne
A joyous smile upon his face.
And though his ears may seem outgrown
He carries them with pride and grace.

He's never cross or quick to carp
A friend in need is he to me.
When human tongues are mean and sharp
My Teddy gives me sympathy.

To him I always bare my soul
He lifts me when I'm feeling low.
And when I brag and miss my goal
He never says, 'I told you so'.

My friends may titter gleefully
And some may tease, but I don't care.
I hope that I will never be
Too old to love my Teddy Bear.

Jeffrey S Forman, *The New York Times*,
October 27 1977

Peter Bull was one of the world's most famous teddy bear lovers and while he had a huge collection of many, his favourite and life-long companion was a little bear who travelled in his pocket. His name was Theodore.

To me he is factual and as real as part of my life as anything I possess. He doesn't remotely resemble a favourite watch or any really inanimate object but I would no more dream of going away without him, even for a night, than flying to the moon. But then I've never really fancied *that*, though I think Theodore might rather like it.

Yet I know that the same thing would happen on the moon as it does in New York, Greece, Hollywood and Nether Wallop, i.e. that the moment I unpack and put Theodore on my bedside table with his friends and props, the strange place becomes a sort of home. I think he's a symbol of unloneliness. He sits there on his haunches (how he *hates* standing up!) reminding me of the happy and unhappy times we've had together, and his funny little face never fails to give me a lift if things are looking a bit black.

Peter Bull, *Bear With Me*, Hutchinson 1969

It must be eleven years now since Kathy went for a walk with the cleaning lady and came home with a Teddy Bear she won in a Friendly Sons of St Patrick raffle they passed.

He is still going strong. He is no longer pink fuzz (or was he blue when he was new?), but grey homespun, and his left arm has a red patch and his glass eyes have long since been swallowed by the current dog. He is a heartbreaking sight. But he is a great traveller, Teddy. He has been to California, Cape Canaveral, and Key West; to Colombia, Cuba, and Kingston, Jamaica; to Connecticut, Cape Cod, and Killooeet Camp. He is a regular Columbus. He has travelled more in his eleven years, as a matter of fact, then I have in thirty-nine or so. I have never travelled parcel post. Teddy has. In a box with a hole punched on the bottom so that he could breathe.

Teddy was not only a good companion and a perfect bed-mate; he was the quickest way to a chambermaid's heart. The sternest of them would push her way into the girls' room, stiffen at the laundry stored in the bidet and mutter "slobs" in Schweitzerdeutsch, only to melt into fondue at the sight of Teddy, snuggling innocently among the rumpled sheets. He turned bed-making into a privilege. He got a ribbon on his neck in Paris, two pillows in Athens and a nightgown made out of a handtowel in Rome.

Violet Weingarten, *You Can Take Them With You*, 1961

A row of teddy bears sitting in a toyshop, all one size, all one price. Yet how different each is from the next. Some look gay, some look sad. Some look stand-offish, some look lovable. And one in particular, that one over there, has a specially endearing expression. Yes, that is the one we would like, please.

Christopher Milne, *The Enchanted Places*

Chapter 4

PARTY BEARS

THE TEDDY BEAR DANCE

When the moon hangs over the rolling downs
That ripple between the seven towns,
When the night is fresh and the air is clear
When the owls hide their eyes and the fire-flies appear,
Then Teddy Bear time is here.
There are Teddy Bears dancing here.

Down sheets and trees and ivy plants
To come to the site of the Teddy Bear dance,
Out from the windows in seven towns
They creep on their way to the rolling downs.
For Teddy Bear time is here.
There are Teddy Bears dancing here.

Out from the beds of their daytime chums
Silently drawn by the beat of the drums,
Out from the houses and out through the gates
Down the dark lanes where they join with their mates.
For Teddy Bear time is here.
There are Teddy Bears dancing here.

Then onto the downs and around they prance
Joining their hands in a frenzy of dance,
Coming together to quadrille in teams
Shedding their stuffing and straining their seams.
For Teddy Bear time is here.
There are Teddy Bears dancing here.

Quicker and quicker the music's beat
Faster and faster the speed of their feet,
Till their eyes shake out and their ears come away
And their fur starts to fly and their paws start to fray,
For Teddy Bear time is here.
There are Teddy Bears dancing here . . .

But when the dawn peeks out over the hill
The drumming stops dead and they all fall quite still:
Knowing it's late and they must get home soon
But caught by the spell of the last of the moon,
For Teddy Bear time was here.
There were Teddy Bears dancing here.

Then laughing and giggling they glue back each bit
They have lost in the night with Teddy Bear spit.
And murmuring and whispering "cheers" to their mates
They rush to their homes and run through the gates.
And scamper up ivy, up sheets and up trees,
And leap through the windows to squeeze, if you please,
Under the covers right down to the ends
Of the innocent beds of their innocent friends,
As if to the world, such a sweet little sight,
They'd been lying there faithful the whole of the night.

But Teddy Bear time was there.
There were Teddy Bears dancing there.

Anon

TEDDY BEARS' PICNIC

If you go down in the woods today
You're sure of a big surprise
If you go down in the woods today
You'd better go in disguise.
For ev'ry Bear that ever there was
Will gather there for certain because,
Today's the day the Teddy Bears have their Picnic.

Picnic time for Teddy Bears,
The little Teddy Bears are having a lovely time today.
Watch them, catch them unawares
And see them picnic on their holiday.
See them gaily gad about,
They love to play and shout, They never have any cares;
At six o'clock their Mummies and Daddies
Will take them home to bed,
Because they're tired little Teddy Bears.

If you go down in the woods today
You'd better not go alone
It's lovely down in the woods today
But safer to stay at home.
For ev'ry Bear that ever there was
Will gather there for certain, because
Today's the day the Teddy Bears have their Picnic.

Music: J E Bratton 1907,
Lyrics: Jimmy Kennedy, Teddy
Bears' Picnic

ACKNOWLEDGEMENTS

The Publishers have made every effort to trace copyright holders. If we have inadvertantly omitted to acknowledge anyone, we should be most grateful if this could be brought to our attention.

Permission to publish *Summoned by Bells* by John Betjeman granted by John Murray (Publishers) Ltd. *The Enchanted Places* by Christopher Milne is published by Methuen, London. Extract from *You Can Take Them With You* by Violet Weingarten. Copyright © 1961, renewed 1989 by Violet Weingarten. Used by permission of Dutton Signet, a division of Penguin Books USA Inc.

The Publishers would like to thank the following photographic libraries for their kind permission to reproduce their photographs:

The Bear Museum, Petersfield, front jacket, p: 1, 7, 8, 9, 11, 12, 15, 16, 19, 21, 24, 27, 28, 33 (left), 39, 40, 42, 44, 47, 48, 50, 51, 54, 57. The Museum of Childhood, Edinburgh p: 41, 61. Sylvia Cordaiy / Eddy Mayhew p: 14, Sylvia Cordaiy / John Periam p: 29, Sylvia Cordaiy / Linda James p: 30, 60, Sylvia Cordaiy / Monika Smith p: 55, Sylvia Cordaiy / John Jones p: 59. Robert Harding Picture Library p:20. Photographic Perspectives / Photograph © Jeremy Neave 1994 (All Teddy Bears and props kindly loaned by Park House Antiques, Stow on the Wold.) back jacket, p: 3, 6, 10, 13, 17, 18, 22, 25, 26, 33 (right), 35, 36, 37, 38, 45, 46, 49, 52, 56, 62, 63.

Thanks also to Pat Rush, Editor of *Hugglets* magazine, for her help and advice.